D0787658

The iguanodon died a long time ago.

A really long time ago.

Bit of a bummer way to start a story,
but hang on.

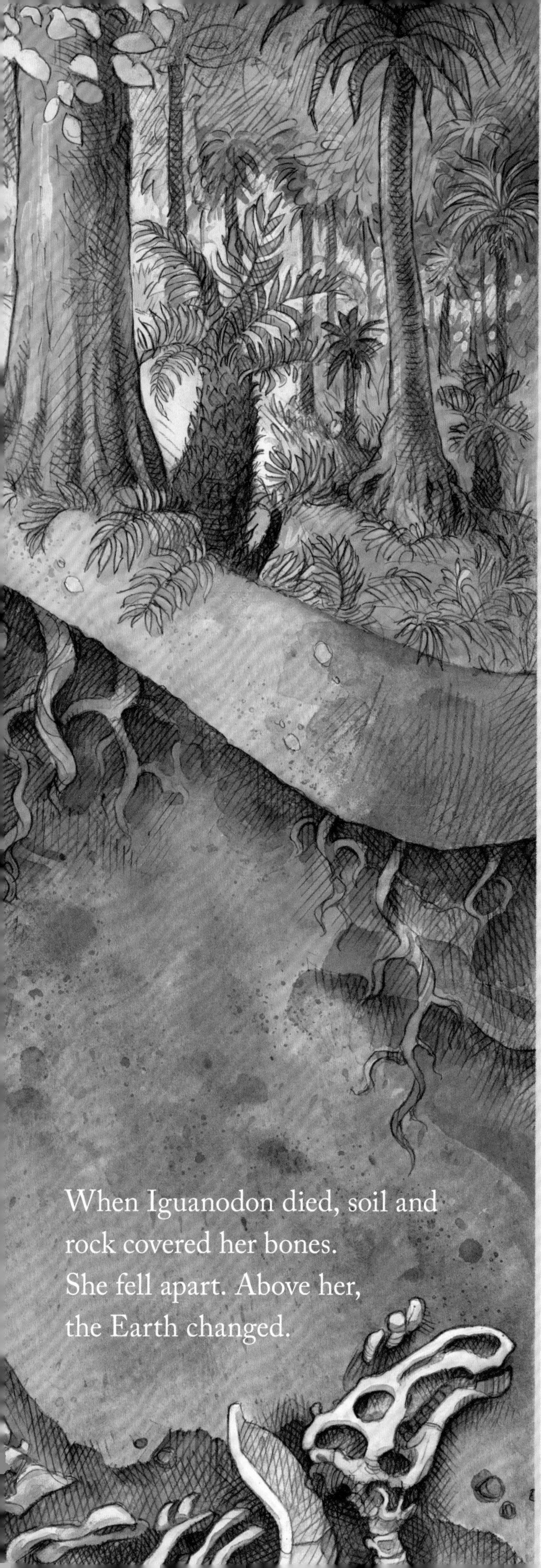

When Iguanodon died, soil and rock covered her bones. She fell apart. Above her, the Earth changed.

Many years later, humans began to explore. They didn't always understand what they found.

When humans stumbled across the remains of animals like Iguanodon, they created stories about them. They thought the bones came from creatures of myth—

Basilisk
Chinese Dragon
龍
European
Dragon

MONSTERS and DRAGONS!

As humans discovered more about the natural world, they began to use experiments and observation to better understand their discoveries.

Today, we call this SCIENCE.

1700 · CHEMISTRY

Science is a process. When humans do science, they create theories—sets of ideas about how the world works. Sometimes the ideas are right. Sometimes they're only part right. And sometimes the ideas *seem* right—even when they're actually wrong.

1785 · GEOLOGY

1859 · NATURAL SELECTION

That's because we may not
have enough information
to know for sure.
Not yet anyway.

1811 · PREHISTORIC REPTILES DISCOVERED & then . . .

Whenever new information
is discovered, an old theory
may be broken down and rebuilt.
And rebuilt. And rebuilt.

Because science is a process
that never ends . . .

. . . but it needs to start somewhere.

SEAN RUBIN
the IGUANODON'S HORN
How Artists and Scientists put a DINOSAUR back together again and again ... and AGAIN

CLARION BOOKS
An Imprint of HarperCollins*Publishers*

One day, a human named Mary Ann Mantell was out walking. She came upon a large tooth, unlike any she had ever seen before, buried in the road.

Mary Ann examined the tooth with her husband, Gideon.

Gideon shared the tooth with other scientists. They thought it may have come from a rhinoceros. Or an enormous iguana.

It didn't.

Gideon named the creature it came from IGUANODON, which means "iguana tooth."

Gideon and Mary Ann found more bones and teeth. They wondered what a *whole* iguanodon looked like.

Reconstructing Iguanodon was like trying to put together an incomplete puzzle. First, the Mantells studied the pieces they had and considered all the evidence. Then they used their imaginations.

Their theory was simple—Iguanodon was an enormous reptile who ate plants. They even made drawings so other people could see their ideas.

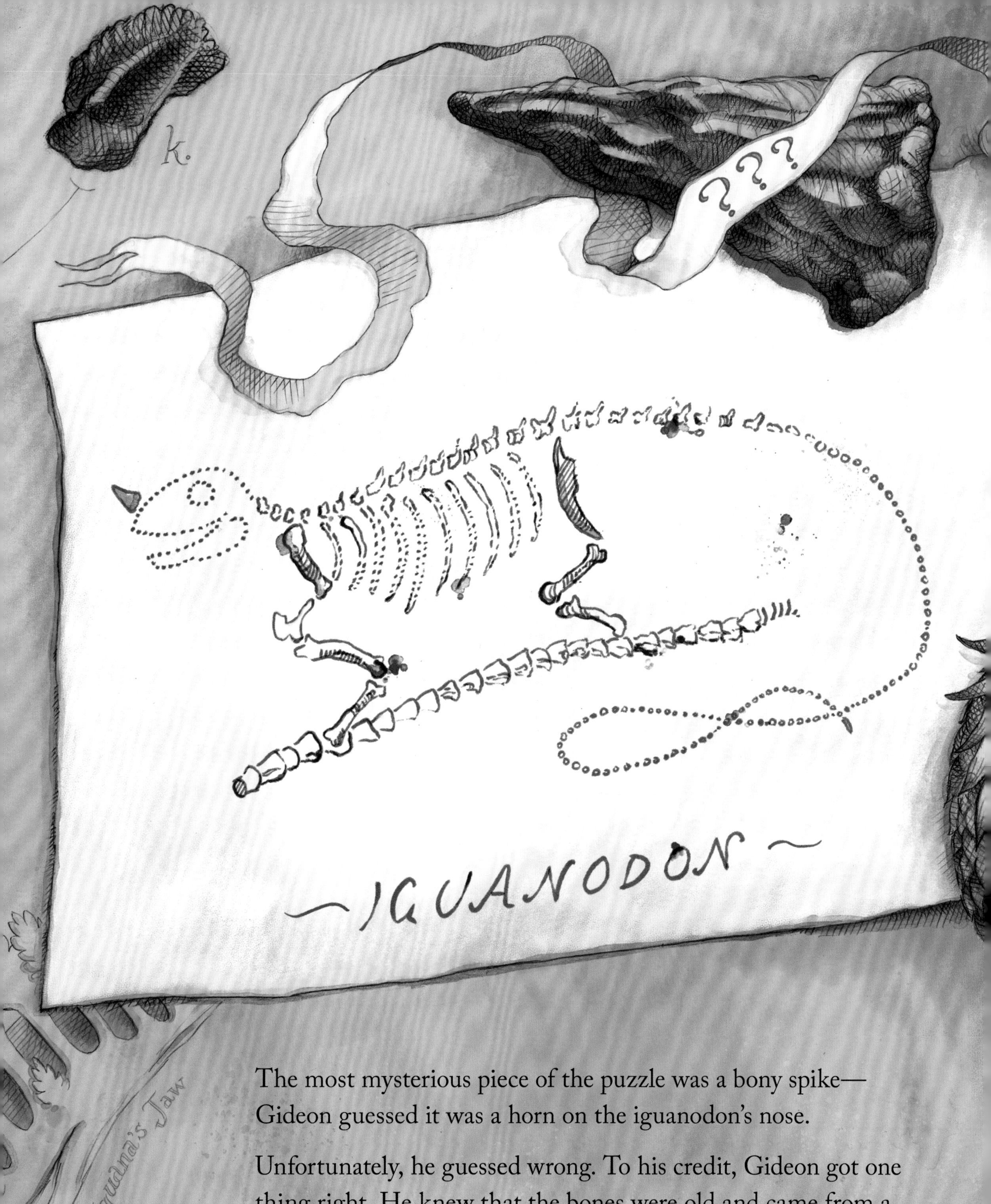

The most mysterious piece of the puzzle was a bony spike—Gideon guessed it was a horn on the iguanodon's nose.

Unfortunately, he guessed wrong. To his credit, Gideon got one thing right. He knew that the bones were old and came from a creature that no one had ever seen before—an animal previously unknown to science.

Iguanodon was a . . .

DINOSAUR!

THE "FIRST" IGUANODON
(ca. 1850)

- Nose horn
- Lizard tongue
- Long dragon tail
- Basically an iguana the size of a whale

Paleontologist: Gideon Mantell

Artist: Mary Ann Woodhouse Mantell

di·no·saur from the ANCIENT GREEK "Terrible Lizard"

I DO look pretty terrible here.

I know what you're thinking. Dinosaurs didn't look anything like this. Especially not Iguanodon. But in 1850, people didn't know that yet.

Remember, at this point, no one had seen a complete dinosaur skeleton. All they could do was make creative guesses. *Really* creative guesses.

The Mantells' work inspired other scientists to look for dinosaur remains. As more fossils were discovered, artists began creating their own reconstructions—drawings, paintings, and even sculptures of what dinosaurs may have looked like.

Because most skeletons were incomplete, artists based their work on animals they already knew—lizards, whales, and, yes, rhinoceroses. Their imaginations took off from there.

They drew dinosaurs as scaly monsters locked in mortal combat, rising out of murky swamps.

These were the first paleoartists—artists who interpret fossils and help us understand how extinct creatures may have appeared and behaved. One of these artists was Benjamin Waterhouse Hawkins.

{ with Year of Discovery }

In 1852, Benjamin created enormous dinosaur sculptures for a special exhibit in London. To celebrate the exhibit, Benjamin invited a group of scientists to a New Year's Eve dinner . . . inside the iguanodon sculpture.

To create his iguanodon, Benjamin worked with Richard Owen, the paleontologist who coined the term "dinosaur." The sculptures were very popular. People came from all over to see them. Unfortunately, they were totally inaccurate.

CRYSTAL PALACE IGUANODON (ca. 1854)

- Nose horn
- Huge claws
- Sharp teeth
- Thick scales
- Walks on all fours
- Slow-moving
- Looks something like a cross between a rhinoceros and an iguana

Paleontologist: Sir Richard Owen

Artist: Benjamin Waterhouse Hawkins

Still, there was reason to be optimistic. Someone was bound to put Iguanodon back together the right way. Eventually. Now that everyone knew dinosaurs existed, even more people were on the lookout for fossils. New information was sure to turn up.

It didn't take long.

In 1878, Belgian coal miners found fossilized iguanodon skeletons one thousand feet below the Earth's surface. Complete skeletons. All the bones!

The scientist who reassembled these skeletons was Louis Dollo.

Louis was working with new clues about Iguanodon, clues that showed the old theories weren't quite right. Because he had complete skeletons, Louis knew that to put them together correctly, he'd have to start from scratch.

First, Louis compared the iguanodon bones to the skeletons of other animals in his museum.

He thought Iguanodon's big hind limbs and smaller forelimbs looked a bit like a giant flightless bird—an emu. But the dinosaur's limbs also reminded him of a kangaroo or a wallaby.

Unfortunately, Louis based his reconstruction more on the wallaby. He guessed wrongly that Iguanodon stood upright on two legs and dragged her tail.

On the plus side, Louis discovered that Iguanodon's so-called nose horn wasn't a nose horn at all: turns out she had two of them, and they attached to her hands.

Iguanodon had thumb-spikes!

Thanks to Louis's work, Iguanodon finally looked nothing like a rhinoceros.

TWENTIETH-CENTURY IGUANODON (ca. 1900–1970)

- Thumb-spikes
- Stands upright
- Drags its tail
- Lizard tongue and lips
- Probably lumbers
- Kind of like a big, bipedal, kangaroo-shaped monitor lizard

Paleontologist: Louis Dollo

Artists: Charles Knight, Rudolph Zallinger

Louis's version of Iguanodon inspired countless artists, including Charles Knight and muralist Rudolph Zallinger. In the following years, paleoartists depicted dinosaurs as lumbering, tail-dragging, cold-blooded reptiles, too dim-witted to avoid extinction. These inaccurate images led some humans to believe—also inaccurately—that dinosaurs were, well, boring.

By the beginning of the twentieth century, dinosaur studies had entered a slump. Fortunately, they didn't stay in a slump for long.

In fact, dinosaurs were about to take flight.

For over a century, dinosaurs were thought to be a lot like reptiles. But by the 1960s, new evidence challenged this assumption.

Crucial information came from trackways—a series of fossilized dinosaur footprints. Modern reptiles are cold-blooded and drag their tails, leaving grooves in the dirt behind them.

When iguanodon trackways were discovered, scientists were surprised to find no tail grooves. While the trackways showed Iguanodon had walked on four legs, she had held her tail even with the ground—like a bird.

Most importantly, there was evidence that dinosaurs lived active lifestyles—that they hunted, foraged, and even migrated. The evidence also suggested that dinosaurs may have been warm-blooded, like mammals or birds.

A scientist named John Ostrom had started to form a new theory—a set of ideas that would completely change the way humans imagined dinosaurs.

What if dinosaurs weren't just like reptiles after all? What if dinosaurs were a lot more like . . .

BIRDS?

Just as Louis had broken down
Gideon's version of Iguanodon,
now John would break down everyone's
version of almost every dinosaur. This was a
huge job. He considered all the information,
old and new, and came up with fresh ideas.

1969 · DEINONYCHUS IDENTIFIED

John's theory was strengthened by the discovery of Deinonychus—a fossilized dinosaur that clearly resembled a bird skeleton.

John asked his friend Bob Bakker—a scientist who was also an artist—to draw a picture of an active, warm-blooded, birdlike dinosaur so that others could understand his ideas.

Not only did dinosaurs act like birds, some even looked like birds.

Humans began to change their minds about dinosaurs. Much of what they thought they knew had been based on incomplete information.

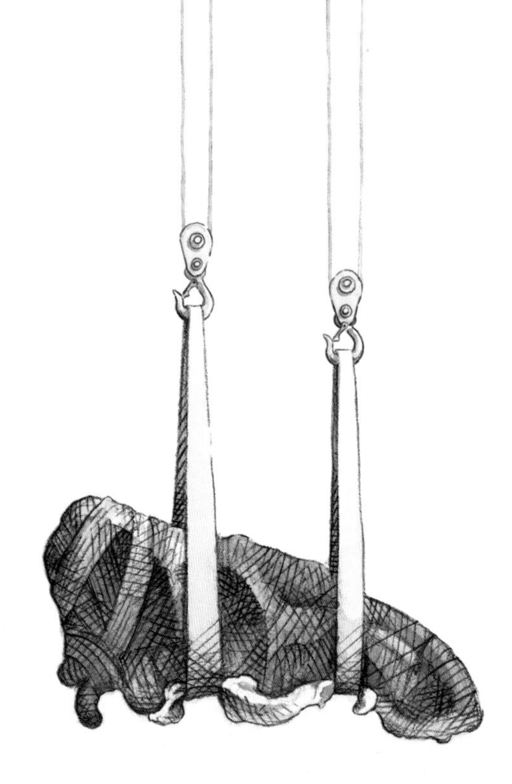

The scientists and artists were determined to fix their mistakes.

Using new information, scientists remounted museum fossils in more birdlike poses, with their tails *up*.

Artists began to draw and paint dinosaurs that looked smart, active, and warm-blooded. Finally.

Some call this time the Dinosaur Renaissance: "renaissance" means rebirth. Dinosaurs *were* reborn—and everyone noticed.

THE RENAISSANCE IGUANODON
(ca. 1970–2000)

- Thumb-spikes
- Typically quadrupedal as adults, may stand on two legs to reach food
- Bipedal when young
- Warm-blooded
- Swan neck, no lips, beak!
- Skin pulled tight over its body, little soft tissue or color

Paleontologist: John Ostrom

Artists: Bob Bakker, John D. Dawson

By the year 2000, scientists could make many informed guesses about Iguanodon. They knew she had a spike on her hand. They thought she usually walked on four legs, although not always. She almost certainly held her tail even with the ground.

Scientists were also pretty sure iguanodons were warm-blooded, lived in groups, built nests, and cared for their young—like birds.

Artists drew Iguanodon according to what scientists thought. But some artists were beginning to wonder why dinosaurs always looked so scrawny, bony, and sort of . . . shrink-wrapped.

Many animals alive today have feathers or fur.

What's more, they have curves, fat, fleshy bits, and other soft tissue that isn't obvious if you're just looking at a skeleton.

Any idea what animal this skull comes from?

the HippoPotamus!

Are you surprised?

Hippos don't look much like their skeletons. That's because most of the hippo isn't bones. A hippo is also made of skin, hair, muscle, and fat.

Extra hippo!

If you tried to draw a hippo without any extras, it might look something like this:

Dinosaurs were no different. Bones can't tell us everything about a dinosaur's shape, and it's important to look at other clues—including fossilized skin, muscle, and even feathers. Once again, artists are studying all the evidence, then letting their imaginations run wild. They're giving dinosaurs more fat and more feathers. And dinosaurs have changed again.

Dinosaurs weren't monsters.
They weren't dragons.
They were . . .

ANIMALS!!!

Like modern animals, they
came in all sorts of colors.

They smelled funny.

Dinosaurs had baggy bits and saggy bits.

They made messes and strange noises.

And they had weird, showy parts that only existed to attract mates.

Are any of these drawings accurate? Unless someone invents a time machine, we'll probably never know for sure.

Still. Any day now . . .

. . . someone could find something new. An artist could figure out another piece of the puzzle.

And what you thought you knew about dinosaurs could change—

—*AGAIN!*

SPECULATIVE IGUANODON
(ca. 2000 onward)

- Thumb-spikes
- More soft tissue
- Quadrupedal, large and muscular
- Forelimb ending in something like a hoof
- Possible keratin quills, or even feathers
- Varied coloring

Inspired by the art of John Conway and Natee Himmapaan

Pages 2–5
Stories about dragons and other fantastic creatures may have begun when ancient peoples discovered large and unusually shaped bones—fossils from dinosaurs and other extinct animals—and sought to make sense of them. Adrienne Mayor at Stanford University has written several books on the growing field of geomythology, the study of how people understood nature before the Scientific Revolution. Considered to have started in the sixteenth century, the Scientific Revolution marks a time when many people began using systematic observation to understand the natural world; previously, people tended to rely on intuition and traditional explanations passed down from their ancestors.

Pages 6–7
Page 6 depicts a nineteenth-century naturalist, in the spirit of Alexander von Humboldt and Charles Darwin, studying unique species on the Galápagos Islands. Page 7 features Mary Anning, a homeschooler who discovered the first marine reptile fossil when she was only eleven years old and whose fossil finds launched the scientific community's interest in dinosaurs. Anning went on to become one of the foremost authorities on marine fossils, and she supported her family through the sale of fossilized seashells by the seashore, as they say. Unfortunately, because Anning was a woman, the men who relied on her work never gave her proper credit.

Pages 8–9
When the iguanodon tooth first appeared in the Mantells' possession, Gideon claimed that his wife, Mary Ann Mantell, had found it. After their marriage ended, Gideon said he had found it. While I depict Mary Ann as the finder, it's possible Gideon may have bought the tooth illegally and given his wife credit to cover his tracks. If anyone doubted the story about Mary Ann, they kept their mouths shut—at that time, no one would impugn the honor of a lady.

Pages 10–11
Gideon showed the tooth to other scientists, including Georges Cuvier, a famous French biologist. It was Cuvier who dismissed the tooth as a rhinoceros's tooth. The artist Albrecht Dürer, who had never seen an actual rhino, drew a beautifully inaccurate rhinoceros based on several scientific misunderstandings, and so Dürer's rhinoceros hovers over this gathering of misunderstanding scientists.

Pages 12–13
The art on these pages was inspired by work by Marie-Denise Villers, Gideon Mantell, and Mary Ann Mantell. Prior to the invention of photography, scientists relied on drawings, etchings, and taxidermy to understand animals. Drawings could be reproduced in books purchased only by those with enough money to afford a library. In short, scientific images were hard to come by.

Pages 16–17
This image owes much to the work of artists Benjamin Waterhouse Hawkins, Josef Kuwasseg, and George Scharf. These creatures look especially strange as not only had these paleoartists never seen a dinosaur, they probably had never seen a real iguana, whale, or rhinoceros either. They relied on often-inaccurate drawings to guide their designs, resulting in the odd, dragon-like creatures depicted here.

Pages 18–19

This image includes references to Victorian-era engravings by Philip Henry Delamotte and drawings by Benjamin Waterhouse Hawkins.

Sir Richard Owen dominated paleontology for much of the nineteenth century. This wasn't always a good thing. Owen disliked Gideon Mantell, and when Mantell sought to correct his depiction of the iguanodon based on new research, Owen blocked Mantell's ideas because they contradicted Owen's work. It is possible that the later tail-dragging version of dinosaurs could have been avoided entirely if not for Owen's lack of integrity.

And if you think the servants in this picture seem a bit short, you're right—footmen were often around 5 foot 6 inches, so they could fit into a standard uniform.

Pages 20–21

At this point, paleoart was no longer confined to scientific books. Depictions of dinosaurs had entered popular culture, and even the young girl in this illustration holds a souvenir iguanodon.

This image is inspired by an illustration in Matthew Digby Wyatt's 1854 *Views of the Crystal Palace and Park, Sydenham.* The Crystal Palace was a great glass exhibition hall originally built to house the appropriately named Great Exhibition (1851). To capture the unique appearance of British Victorians looking at cool stuff, as seen from behind, I am indebted to countless postcards from the Exhibition depicting just that.

Pages 22–23

We know what this scene looked like because artist Gustave Lavalette created several drawings of the skeletons in their "death poses," before the fossils were removed from the mine. The original position of the bones was later recreated in the Museum of Natural Sciences in Brussels, Belgium.

Pages 24–25

By this time, photography had become more common, and this illustration was created referencing photos of the actual reconstruction process. Louis Dollo and his assistants rebuilt the iguanodon skeletons in a church, the only building with ceilings high enough to accommodate the upright poses.

Pages 26–27

The mural depicted here is an imagined combination of Rudolph Zallinger's *The Age of Reptiles* in the Yale Peabody Museum and Charles Knight's paintings of brontosauruses and stegosauruses. These images inspired generations of toy designs and came to define the public's image of dinosaurs through the early 1990s.

Page 30

This illustration riffs on iconic images created by cinematographer Dean Cundey for the opening of the film *Jurassic Park* (1993). In some ways, ideas about dinosaurs really only changed for the public when kids' toys changed, and that happened after the release of this iconic movie. (The author would like to add that he still has many of his *Jurassic Park* action figures, and his kids now play with them.)

Page 31
Bob Bakker was a student of John Ostrom. Bakker is shown creating his famous rendering of Deinonychus, which appeared in John Ostrom's "Osteology of *Deinonychus antirrhopus*, an Unusual Theropod from the Lower Cretaceous of Montana," first published in July 1969 in the *Bulletin of the Peabody Museum*, Volume 30. Several roadrunners look on.

Page 32–33
With *Jurassic Park* having informed the public about "new" dinosaur models, some museums began the long, difficult process of updating their fossil mounts. The Dinosaur Renaissance inspired *Jurassic Park*, and the movie's popularity led to more visitors at museums and more students studying paleontology. Science influences art, and then art influences science.

Page 36–41
The ideas and illustrations on these pages are greatly indebted to the collected work of Darren Naish and other paleoartists in *All Yesterdays: Unique and Speculative Views of Dinosaurs and Other Prehistoric Animals*. William, an ancient Egyptian blue hippo figurine from the Metropolitan Museum of Art in New York, makes an appearance as an early example of the human desire to depict animals through art.

The extraordinary paleoartists working today are too numerous to list in their entirety, but in creating the designs and color schemes for these final reconstructions, I was inspired by the artwork of Luis V. Rey, Gabriel Ugueto, Lindsey Wakefield, and a variety of anonymous social media accounts, most notably @dinosaur_models_artworks.

As this book incorporates images and artistic styles first created by many other artists, I tried to acknowledge many sources directly throughout the text and endnotes. I would also like to acknowledge Steven Bellettini, Brian Floca, James Gurney, Natee Himmapaan, Dr. Kathryn Hoppe, Dr. Jack Horner, Sarah Miller, Dr. Jeff Richmond-Moll, and Brian Selznick for their insight and inspiration. Any remaining inaccuracies are entirely on me. Thanks especially to Anne Hoppe, Whitney Leader-Picone, Marietta Zacker, and Jen Keenan, each of whom believed in this book immediately. Thank you to Gordon Scharf, my studio mate of three years, for inspiring the narrator's unique voice; to Sammy, Charlie, and Dax for (not) providing too many distractions; to my parents for always buying me as many dinosaur books and toys as I wanted; and finally, to my wife, Dr. Lucy Guarnera, for tirelessly explaining what science can and cannot do, especially on behalf of the most vulnerable.

Biscuit Run, Virginia
Advent 2022

To Lucy—*clever girl*

Clarion Books is an imprint of HarperCollins Publishers.

The Iguanodon's Horn: How Artists and Scientists Put a Dinosaur Back Together Again and Again and Again

For information address HarperCollins Children's Books, a division of HarperCollins Publishers, 195 Broadway, New York, NY 10007.
www.harpercollinschildrens.com

Library of Congress Control Number: 2023933841
ISBN 978-0-06-323921-0

This book is set in Adobe Caslon, a revival of an eighteenth-century typeface created by William Caslon, which is similar to typefaces used in scientific publications during the early days of paleontology. In 1990, Caslon's original type was dug up and reconstructed by designer Carol Twombly.

To create the art for this book, the artist drew with HB and 2B Staedler graphite pencils on 400-weight Strathmore bristol board. The line art was scanned into a Mac mini, and he added digital watercolor and pastel using Adobe Photoshop on a Wacom Cintiq 32 tablet. Additional texture in the form of physical watercolor washes and paint splatters was also scanned into Photoshop and incorporated into the art. The artist is not entirely sure what to call all this, but "digital collage" sort of makes sense.

Design by Whitney Leader-Picone
23 24 25 26 27 RTLO 10 9 8 7 6 5 4 3 2 1

First Edition